How can a magnet move another object without touching it?

OpenSciEd

MIDDLE SCHOOL SCIENCE

How can a magnet move another object without touching it?

Magnets

OpenSciEd

How can a magnet move another object without touching it?

Magnets

The development of this material was supported by the Bill & Melinda Gates Foundation, Carnegie Corporation of New York, Charles and Lynn Schusterman Family Foundation, and William and Flora Hewlett Foundation.

To download the handouts, slides, and other unit resources, visit OpenSciEd.org and complete the free registration. Accessible versions of the Student Edition, for students with disabilities, are available in the NiMAC Inventory.

TABLE OF CONTENTS

Student Procedures

Lesson 1: What causes a speaker to vibrate?

What's going on inside a speaker?

For anything that produces sound, there has to be a force that is applied to the sound source to deform or bend it to get it to vibrate. We identified what that force was for many instruments, but we never identified what the force was in the speaker to make it vibrate.

Turn and talk

1. How could we investigate what is causing the speaker to vibrate?

In your notebook

2. Make a Notice and Wonder chart in a new page in your notebook, like the example below.

What I noticed	What I wonder

3. Watch the video of the speaker in slow motion that your teacher will show to the class. Record your observations in the left column under the heading "What I noticed". Record your initial questions about what you observed in the right column under the heading "What I wonder." Be prepared to share your ideas with the class.

Modeling the speaker system based on observations

In your notebook

4. On a new 2-page spread, develop a model on the left page of a working speaker based on what you observed from the dissected speaker. Leave the right page blank.
    - Leave space at the top of the page to write in the title later.
    - Include the important parts of the speaker.
    - Show how these parts work together in this system.
    - Show how the parts working together cause the speaker to vibrate.

With your class

5. Make a chart with your class of the important parts of the speaker, why each part is important, and questions we have about each part.

6. How do you think building a homemade speaker could give us more clues about what is producing the force or forces that cause speakers (in general) to vibrate?

With your group

7. Build your speaker.
    a. Wind the wire into a coil. Wrap each end of the wire around your coil 2 times to hold the coil together. Leave at least 3 inches of wire loose on each end.
    b. Use the sandpaper to rub off the enamel on the ends of the wire. You should see copper wire underneath.
    c. Tape the coil to the bottom of your cup.

8. Test your speaker.
    a. Attach the two free ends of wire on your cup speaker to the two alligator clips at the sound testing station.
    b. Put the magnets near the wire to see what happens.

In your notebook

9. On the notebook page to the right of your store-bought speaker model, develop a model of the homemade speaker.
    - Include the important parts and label them.
    - Show how these parts work together in this system.
    - Show what causes the homemade speaker to vibrate and make sound.

Title the full spread "Initial Models of a Speaker System." Label the left page "Store-bought" and the right page "Homemade."

If you finish early, discuss with your group what similarities and differences you see between the two speakers.

With your class

10. Discuss the following prompts:
    - How did building the speaker help you understand the important parts of the system better?
    - What questions came up for you as you built the speaker and compared it to the store-bought speaker?
    - How do you think the magnet and wire work together without touching each other?

Initial consensus model in a Scientists Circle

With your class

11. With your class, you will form a Scientists Circle and revisit the norms for how to work together and learn together in science class.

Then you will develop a record of what we agree on and where we have competing ideas across the initial models. As your classmates share their models, mark on your model:

- If something is similar, place a small check mark (✓) near what is similar on your own model.
- If something is different, place a question mark (?) near what is different on your own model.

Turn and talk

12. Where else have we seen these same kinds of parts in a different system making something move?

With your class

13. We will make a public record of our ideas in a chart like the one below.

What is the system? What parts move?	How does it move?	Are the parts of the system touching?

Navigation: Introducing the unit question

Close-up of the store-bought speaker.

Turn and talk

14. In the photo above, compare the thickness of the coil to the space where the coil sits. Are the magnet and the wire touching inside the store-bought speaker?

In your notebook

15. Turn back to the page in your science notebook that you reserved for the unit title. Record the unit title: *How can a magnet move another object without touching it?*

Using a cause-and-effect framework to build our Driving Question Board

On your own

16. What questions do you have about
    - the speaker system?
    - specific parts of the speaker (like magnets)?
    - other related phenomena that you believe would be similar to the speaker?

On your own, choose 1 or 2 questions to share with your class. Write each question on a sticky note or card. Write the question in dark marker and large enough for others to read. Be prepared to post your questions to our Driving Question Board.

A cause-and-effect relationship can usually be described with a sentence that looks like this:
When we ___*change to the system*___, we observe ___*effect on the system*___.
As an example: When we *drop a cell phone with enough force*, we will observe *the screen shatter*.

Turn and talk

17. Share an example of a cause-and-effect relationship in your everyday life by filling in the blanks in the sentence above.

Stop and Jot

18. Title a new notebook page "Cause-and-Effect Observations." On your own, stop and jot some ideas about what cause-effect statement we can make about the speaker right now. Be ready to share your ideas with the class.

There are a lot of ways that scientists ask questions about cause-and-effect relationships in systems. We could ask a question about the effect. We could ask a question about the change we need to make to get the effect. We could ask a how or why question about a cause-and-effect relationship we are pretty sure about. To capture all these questions, we can extend the cause-effect sentence like this:
When we ___*change*___, we observe ___*effect*___ because ___*how or why explanation*___.

On your own

19. Write 1-2 additional questions (one per note card) for the DQB about a cause-and-effect relationship in the speaker system. Be prepared to post your questions to our Driving Question Board.

Scientists Circle

20. Take out your note cards with questions. Bring those with you to the Scientists Circle along with your science notebook.
    - The first student reads a question aloud to the class then posts it on the DQB.
    - Students who are listening should raise their hands if they have a question that relates to the question that was just read aloud.
    - The first student selects the next student whose hand is raised.
    - The second student reads a question, says why or how it relates, and posts it near the question it most relates to on the DQB.
    - That student selects the next student.
    - Continue until everyone has at least one question on the DQB.

Ideas for future investigations

In your notebook

21. What kinds of investigations could we do and/or what additional sources of data might we need to figure out the answers to our questions? Add your ideas to a new notebook page titled "Ideas for Future Investigations and Data We Need." Be prepared to share these with the class.

Lesson 2: What can a magnet pull or push without touching?

Navigation

For anything that produces sound, there has to be a force that is applied to the sound source to deform or bend it to get it to vibrate. We identified what that force was for many instruments, but we never identified what the force was in the speaker to make it vibrate.

With your class

1. We suspected that the magnet was pushing and pulling on something. Why? What do we already know about what magnets can push and pull on?

Turn and talk

2. What do you predict will happen when we put the magnet up close to the metal wire without letting it touch?

Investigating Magnets

With your group

3. Record the investigation question in your notebook: *What can a magnet pull or push without touching?* Add the handout below the question. Follow the instructions in the handout. When you are finished with the investigation, respond to the discussion questions.

Building Understandings

Turn and talk

4. When we bring a magnet close to another magnet, do you notice forces on only one of the magnets or is it happening on both of them?

5. Consider these questions first on your own and then with a partner. Be ready to share with the class.
    - What can this table tell us about the metals that interacted with the magnet to produce forces?
    - Can this help explain why there wasn't any force between the coil of wire and the magnet?

6. A compass produces both attractive and repulsive forces when it interacts with the magnet. Is the compass needle a permanent magnet, or is it just made of a metal that interacts with magnets, like iron? Why do you think that?

Tracking Our Ideas

On your own

7. You have developed a couple of really important ideas today. Let's keep track of them in the Progress Tracker in your notebook. Record the lesson question on the left side of the table and record your ideas on the right side of the table. You can use any combination of words and pictures.

Navigation and Exit Ticket

On your own

8. Respond to the following as an exit ticket: Is there something we could do to cause the coil to interact with the magnet?

Making Predictions

Turn and talk

9. What will be the effect on the coil and the magnet when we connect the coil of wire to a battery?

In your notebook

10. Record the investigation question. Underneath, make an if-then prediction about the effect on the coil and the magnet when we connect the coil of wire to a battery.

Investigating the Coil and the Magnet

In your group

11. Get a battery, coil of wire, two pieces of tape, and a magnet.

12. Connect one end of the coil to an end of the battery. Connect the other end of the coil to the other end of the battery.

13. Use the magnet to try to move the coil without touching it.

CAUTION: The coil can get hot. Disconnect the coil of wire from the battery when you are not actively testing something.

When something moves, we say it has *kinetic energy*. That energy has to come from somewhere. When energy moves, we say it is transferred.

Turn and talk

14. Do magnetic forces always transfer energy?

15. What evidence would we look for to determine if energy had been transferred to or from an object in the system?

In your group

16. Use the handout that your teacher gives you to complete the investigation.

Tracking Our Ideas

Own your own

17. You've developed a couple of really important ideas today. Record your ideas on the right side of the Progress Tracker table in your notebook. You can use any combination of words and pictures.

Navigation and Exit Ticket

We saw the effect of the force as movement. We also remembered from the *Contact Forces* and *Sound* units that when something moves it has kinetic energy.

Turn and talk

18. Where do you think that energy came from if the parts of the speaker are not touching?

On your own

19. Is the coil of wire a magnet? Why or why not? Use at least one cause-effect relationship to justify your response.

Lesson 3: How does energy transfer between things that are not touching?

Navigation: Making Predictions

Turn and talk

1. Discuss the question: How does energy move through the system if the parts aren't touching?

In your notebook

2. Record the lesson question. Below, jot down your initial ideas about what could be happening in the space between the magnets.

Turn and talk

3. Work with a partner to come up with two cause-effect relationships that you predict will tell us something about whether our explanation is true.

In your notebook

4. On a new page in your notebook, write this question: What cause-effect relationship(s) do we predict we will observe if this how or why explanation is true?

Turn and talk

5. Work with a partner to come up with two cause-effect statements.

Develop Hypotheses in Scientists Circle

With your class

6. With your class, you will develop consensus hypotheses in the Scientists Circle.
    - Listen to the ideas of your classmates, paying attention to similarities and differences.
    - Record the class consensus hypotheses in your notebook.

Whole-class Investigations

Turn and talk

7. With a partner, brainstorm ways we could test our cause-effect relationships to establish whether air is involved in causing energy transfer through magnetic forces.

With your class

8. Put *Blocking Air Investigation* into your notebook. Record the variables for the investigation.

With your class

9. Record the investigation results and discuss whether they support the theory that air is involved in the transfer of energy through magnetic forces.

In your notebook

10. Record your observations from the video that your teacher shows you in your notebook. Label the page "Removing the Air."
    - What happens to the music in the video?
    - What happens to the magnetic forces in the video?

Turn and talk

11. Discuss the question: Did we observe in the demonstration or the video any of the predicted cause-and-effect relationships that would support this theory?

Navigation

Turn and talk

12. Discuss these questions:
    - How might we be able to collect evidence to help us understand what is going on in that space between two magnets?
    - What could be happening in the space between the magnets when they pull each other or push each other away if energy is <u>not</u> transferring through air?

Lesson 4: What can we figure out about the invisible space around a magnet?

Exploring the Space Around a Magnet

We have been wondering about this invisible space around a magnet and we want to know about this space and how and why it affects some objects. Now we will use two magnets to find out more.

With your group

1. Get the two magnets from your teacher. If a group of you are sharing the magnets, be sure everyone gets a chance to explore the space around the magnet. Use the two magnets to feel the space around the magnets that is interacting with objects. Try different orientations of the magnets. Notice when they are attracting and when they are repelling each other.
    a. What can you feel?
    b. How does this help you figure out more about this invisible space?

In your notebook

2. We call this space around a magnet a *magnetic field*.
    a. Write the lesson question in your notebook.
    b. Write this new term in your notebook below the lesson question. We will add our definition as we figure out more.

Investigating the Field Around a Magnet

We need a better way to visualize this space--the magnetic field. What if we had lots of tiny pieces of iron and put them near a magnet in the magnetic field?

In your notebook

3. On a new two-page spread in your notebook do these things
    a. Write the lesson question: "What can we figure out about the invisible space around a magnet?"
    b. On the left page, write "Observations of the magnetic field".
    c. Sketch a bar magnet on this page for recording observations.
    d. On the right page, write "Questions about the magnetic field".

On your own

4. Watch the demonstration. As you watch, add to your sketch and record ideas you figure out that will help you understand more about the magnetic field.

WIth your class

5. Discuss these questions with your class.
   a. What did you figure out using the iron filings?
   b. Are the patterns you see the same patterns at each end of the bar magnet?
   c. What have you learned about forces and energy that helps you understand more about the magnetic field?
   d. Where do you think the force is coming from?
   e. Do you think the force is the same at all points around the magnet?
   f. What new questions do you have about magnets or this space around a magnet that we don't quite understand?

Making Sense of Magnetic Fields

With your class

6. Share what you figured out about the magnetic field by sharing your observations with your class. Be sure to give evidence for what you share and what you think your observations mean.

In your notebook

7. Develop a working definition of **magnetic field** with your class.
   a. Use the ideas that you and your class have figured out about the magnetic field.
   b. Find the term, magnetic field, that you wrote in your notebook. Write your definition and leave space below the definition to add more things you figure out.
   c. Share your definition as your teacher directs.

With your class

8. Share your definition for magnetic field as your teacher directs.

Navigate: What do we still need to know about a magnetic field?

The demonstration with iron filings may have made you wonder and ask more questions about the magnetic field. Look back in your notebook at questions you had.

With your group

9. Use the index cards your teacher gives your team and decide on questions your team has about the magnetic field. Write each question on a separate index card. As you write your questions, consider these points:
   a. Is the question testable in our classroom? If not, decide on a different question.
   b. Did everyone in your group contribute to your set of questions? If not, ask each member of your group what questions they had about magnetic fields. If their ideas are not represented and their ideas are testable, write the new question on an index card.

With your class

10. Work with your teacher and the class to add questions to the Driving Question Board (DQB).
   a. Label the new categories of questions your class made for the DQB.
   b. What object can you use to help you learn more about the direction of forces in a magnetic field?

Navigation

We will begin to investigate how forces in the magnetic field compare in different places in the space around a magnet by tackling one of our categories of questions on the DQB.

Getting Oriented to the System

A compass needle is a permanent magnet on a pin. The pin allows the permanent magnet (the arrows) to rotate so that it can line up with the direction of the forces in a magnetic field.

Turn and talk

11. Why is it important to keep track of which end of the magnet is which?

Making Predictions

In your notebook

12. *Which way would the N pole of the compass needle (the red end) point if we moved it to other locations in the magnetic field?* Record your predictions in your notebook by drawing the direction the needle would be facing when moved to each location shown below (B, C, D, and E).

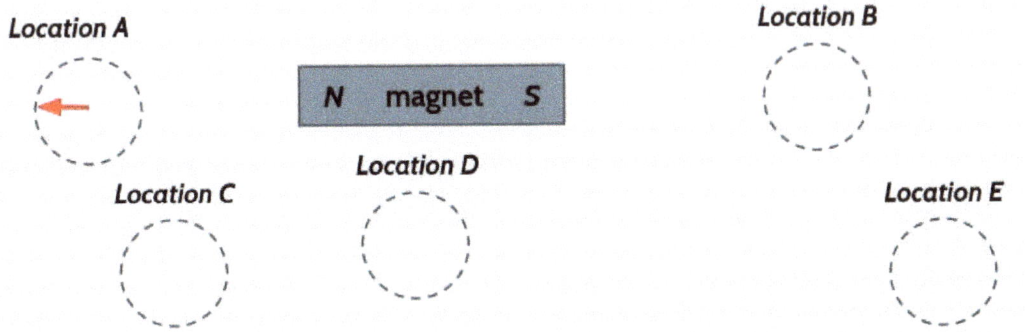

With your class

13. Share your predictions as your teacher directs.

Testing the Field

To collect evidence that helps us answer our questions, we are going to use compasses much like we used the iron filings.

With your group

14. Steps to Take: Getting Ready
    a. Gather your materials. You will need these items:
        - 1 bar magnet
        - 4 compasses
        - 1 piece of white paper

    b. Arrange your supplies so that the magnet is sitting on the white paper.

15. Steps to Take: Testing the Field
    a. Place your compasses all around your magnet.
    b. Sketch the orientation of the compass needle at each location. There is no need to draw the whole compass.
    c. Move the compasses to different locations around the magnet and sketch the orientation of the needle at each location.
    d. Repeat this process until you have a map of the field around the magnet.
16. Discuss these questions with your group.
    a. How do the directions of the forces in the field around a magnet change?
    b. How would you describe the shape of the magnetic field around the magnet?

In your notebook

17. Use words and pictures to add any new ideas you have figured out to your Progress Tracker.
18. Record and respond to these questions in your notebook:
    a. How do the directions of the forces in the field around a magnet change?
    b. How would you describe the shape of the magnetic field around the magnet?

WIth your class

19. Discuss these questions with your class:
    a. How do the directions of the forces in the field around a magnet change?
    b. How would you describe the shape of magnetic field around the magnet?

20. Discuss these questions with your class:
    a. Do you think permanent magnets are the only things that have fields?
    b. How can you find out?
    c. Name at least one place at home you think you will find evidence for a magnetic field.
    d. Name at least one place at home you think you will *not* find evidence for a magnetic field.

Home learning

21. Take the compass home and put it near a variety of objects or devices you have at home. Keep a list of objects or devices you find that have a magnetic field and objects or devices that do not have a magnetic field. Take photographs or video clips of one of the most surprising places you find to share with the class.

Sharing Our Home Learning

Turn and talk

22. Share what you discovered from taking a compass home to look for magnetic fields by discussing these questions:

- What did you find that has a magnetic field?
- How did you test it?
- How do you know it has a magnetic field?
- What do many of these have in common?
- What did you find that does *not* have a magnetic field?
- How did you test it?
- How do you know it does not have a magnetic field?

WIth your class

23. Share your findings with your class as your teacher directs.

We think electricity might have something to do with a magnetic field because of what we saw had magnetic fields in our everyday lives. Was there anything in our speakers that might have had electricity running through it?

Turn and talk

24. What could we use to test our coil of wire?

Testing the Electromagnet

Turn and talk

25. Do you think your coil of wire, connected to a battery, will have a magnetic field? Why or why not?

In your notebook

26. Title a new page in your notebook "My Magnetic Field Predictions" and predict how a compass needle will point at various locations around a coil of wire.

Testing the Magnetic Field Around a Coil of Wire

With your group

27. Follow the instructions for testing your predictions about the magnetic field around a coil of wire.

   a. Lay the coil flat on your desktop.
   b. Connect wire leads with alligator clips to the wires of the coil.
   c. Connect one alligator clip to the battery.
   d. Place small compasses around the coil. Be sure to place them along the coil and at the ends of the coil. Place one compass inside of the coil.
   e. Sketch your arrangement in your notebook. Draw small circles for the compasses but <u>do not</u> draw the compass needle yet.
   f. When everyone in the group is watching, touch the loose wire to the remaining terminal of the battery.
   g. Draw arrows in the compasses to show the direction that the compass needle points when both ends of the coil are connected to the battery.

In your notebook

Safety Precautions

Do not leave the coil connected to the battery for an extended amount of time. The coil will heat up. Leave it disconnected and simply touch the lead to the battery to check the position of the compass needles. You may have to do this several times.

Revising Our Working Definition of a Magnetic Field

It seems we have learned some important things about a magnet field since we first started our working definition. Find your working definition in your notebook and read what you wrote.

In your notebook

28. Our working definition is similar to this: A **magnetic field** is a region of space around a magnet that appears to affect the behavior of certain objects. The objects experience a force when they are in the field. What new things can you add to your working definition?

Turn and talk

29. What would we need in order to see the magnetic field in more detail?

Navigation

With your class

30. What were some of your ideas about how to gather more evidence and details about the magnetic field everywhere around a magnet or a coil of wire?

Using a Computer Simulation to Model the Field

Scientists use computers to reproduce the behavior of a system so that they can see the effects on the system if there is a change to any part of it. We will use an interactive to reproduce the behavior of a system with a magnet and a coil, like in the speaker. We can use the interactive to explore how the magnetic field responds to changes in other parts of the system.

WIth your group

31. Follow your teacher's directions for exploring the magnetic field with the computer interactive.

Add to Your Progress Tracker

In your notebook

32. Add to your Progress Tracker what you figured out using the computer interactive.

Turn and talk

33. How is the speaker system different from the diagram on the slide?

With your class

34. Look at the diagrams on the next two slides and describe how a compass would respond in these situations:
    - when placed next to the N end of a magnet
    - when placed next to the S end of a magnet

On your own

35. What would happen to a compass in the middle of a two-magnet system? How does it know which arrow to follow at this point in space? What happens when you have 2 magnets that are close? How would the magnetic field respond? Why? Use words or diagrams to illustrate the effect you predict. Don't forget a justification for their prediction.

Lesson 5: How does the magnetic field change when we add another magnet to the system?

Magnetic Fields

Turn and talk

1. How does the magnetic field change when we add another magnet to the system?

In your notebook

2. On a new page of your notebook, record a prediction under the lesson question: *How does the magnetic field change when we add another magnet to the system?* You can use words or diagrams to illustrate the effect that you predict on the magnetic field.

Setting Up Your Science Notebook

In your notebook

3. Turn to a new 2-page spread in your notebook. Across the top, label the pages "Magnetic Fields Between a Magnet and a Coil of Wire." Below, label the left page "Attractive forces" and the right page "Repulsive forces." Then draw a horizontal line across the middle of the pages and label the rows "Big gap" and "Small gap."

Why investigate the size of the gap?

Turn and talk

4. Why do we care about the size of the gap between the coil and the magnet?

Using the Interactive

With your group

5. Add a magnet and a coil to the interactive. Don't forget to connect the coil to a battery!

6. Simulate attractive forces between the magnet and the coil with a big gap between them and record the field that you observe in your notebook at the top of the left page.

7. Drag the magnet and coil around to make the size of the gap smaller and record your observations on the bottom of the left page.

8. Do the same thing for repulsive forces (big gap and small gap) and record your observations on the right page.

Coming to Consensus

With your class

9. Come to consensus around how to represent magnetic fields for attractive and repulsive forces between a magnet and a coil of wire. As your classmates share their models, make these marks:
   - If something is similar, place a small check mark (✓) near what is similar on your own model.
   - If something is different, place a question mark (?) near what is different on your own model.

Attractive vs. Repulsive Forces

Turn and talk

10. How would you describe the differences between attractive and repulsive magnetic fields?
11. What happened to the magnetic fields when we pushed the magnet and the coil closer together?

Navigation and Exit Ticket

On your own

12. Without a compass or an interactive to help us map the magnetic field, what effect, if any, do you predict we will be able to observe in real life... when we move magnets closer together
    - if the forces are attractive? Use a "when we...we will observe" sentence to write your prediction.
    - if the forces are repulsive? Use a "when we...we will observe" sentence to write your prediction.

Lesson 6: How can we use magnetic fields to explain interactions at a distance between the magnet and the coil?

Navigation

Open your notebook to the page where you have your original models of the speaker system.

With your class

1. Discuss these questions with your class:
    - When we decided to investigate the magnet and the coil in this set of lessons, what were we trying to explain?
    - What were some of the surprising things we have figured out when we started to investigate the speaker and its different parts?
    - What were some of the surprising things we've figured out when we started investigating the space around the magnet and the coil or wire?

Tracing Cause-and-Effect Relationships

In your notebook

2. Look back in your notebook for ideas about cause-and-effect relationships we have uncovered in our last five lessons.

With your group

3. Work with your group to complete the Cause-Effect table with changes we made to the system and effects we observed when we made those changes.

With your class

4. Work with your class to make a public record of our ideas.

Turn and talk

5. With a partner, discuss which of these cause-and-effect relationships we need to explain the interactions between a coil and a magnet that are not touching?

Making Models

We want to capture what we figured out about cause and effect in our models, so we may need to draw the system several times in order to capture all the important relationships.

In your notebook

6. Individually, record your initial thinking about a revised model of the speaker system.

With your group

7. Work with your group to refine a model of the system that captures what we've figured out about cause-and-effect relationships.
    - Be sure to include all the important parts of the system.
    - Draw the system multiple times to represent all the important relationships.

Navigation and Exit Ticket

On your own

8. What do you think is missing from your model that we need in order to explain how a speaker works?

Compare Models in a Gallery Walk

With your group

9. Visit 2-3 models from other groups and use your notebook to record
    - one idea you see used in each model that may be useful for your class to use.
    - one difference you see where you can argue that your group's model represents the concept or part of the system better.

Build a Consensus Model in a Scientists Circle

In your notebook

10. Find a new two-page spread in your Progress Tracker and draw a three-box tracker as described by your teacher.

Scientists Circle

11. Work with your class to build a consensus model of the speaker system.

Return to the DQB

With your class

12. For answered questions, on the same card as the question, record an answer.

13. Move the answered questions to a different section of the DQB ("Questions we have answered").

14. Add any new questions to the Driving Question Board.

Midpoint Assessment

In the unit so far, we have identified how the parts in a speaker work together to make the speaker move. There is another system that also produces sound, and it has some of the same parts in it that the speaker does. This new system is a doorbell. It is powered by a battery and has copper wire coils in it too. Let's see if using our approaches for thinking about systems and cause-and-effect relationships can help us figure out how this system produces sounds.

15. Complete the midpoint assessment according to your teacher's directions.

Identifying Gaps in What We Know

When we turn music up loud, we sometimes observe nearby windows moving. We can say that the speaker pushes the air with a force, and the particles in the air collide with neighboring air particles, which puts a force on that other air, and eventually air collides with the window. Or we can say that the sound wave transferred energy from the speaker across the air to the window. These are two ways of thinking about the same interaction, and they are both correct.

Turn and talk

16. Where in the speaker system do we have evidence of energy transfer because something starts moving or stops moving?

With a partner

17. Where do we know energy is being transferred between parts of the system? Your teacher will pass out small sticky notes shaped like arrows. Spend some time with a partner (or in table groups) identifying places on the model in your Progress Trackers where you think energy is transferring. Label those with an arrow sticky. Be prepared to share your ideas and justify them.

Exit Ticket

18. Imagine we have two permanent magnets. What changes do we need to make to a system with two magnets to get the effect we would expect if energy transfers
- when the forces are repulsive?
- when the forces are attractive?

Use cause-effect sentences to write your predictions.

Lesson 7: How does changing the distance between two magnets affect the amount of energy transferred out of the field?

Navigation

The results of our explorations with the simulation led us to start to question **what else** might be happening in the field right between the ends of two magnets when we move them closer together or further apart.

Turn and talk

1. Turn and talk to your partner and discuss these questions:
    - What did the simulation results show us that lead us to question this?
    - What other changes do you think we would be able to detect between two real magnets as we move them closer together or further apart?

With your class

2. Debrief your conversation with your class according to your teacher's directions.

Turn and talk

3. Look at the equipment setup of a stationary magnet and a magnet attached to a cart. The magnets are oriented so that they will produce repulsive forces on each other. Discuss these questions:
    - What would you expect to see happen after you bring the cart with the magnet on it close to the magnet on the brick and then release it?
    - Why?

Plan the Investigation

It is often useful to create a hypothesis before planning an investigation. This includes developing a proposed mechanism and a cause-and-effect relationship that you predict you will see if the proposed mechanism is true. You also need to decide what sort of data you need to collect to support or refute the cause-and-effect relationship and thus support or refute the proposed mechanism.

In your notebook

4. Add the two handouts your teacher gives you to your notebook according to your teacher's directions.

With your group

5. With your group, frame a hypothesis about what will happen to the energy transferred out of the field when we push the magnets together and then let go. Remember to include an unobservable mechanism and an observable cause-effect relationship.

Carry Out the Investigation

With your group

6. Carry out your investigation with your group as follows:
    - Read through the procedure with your group.
    - Identify the variables in this investigation and record those on your first handout.
    - Determine the different conditions you want to test and how many trials you want to run for each condition. Make a table on *Unknown material with identifier: df.l6.ho2*
    - Check in your completed investigation plan.
    - Once approved, go to your designated lab station to carry out your investigation plan.

Make Sense of the Data

On your own

7. On your own, answer the questions on *Making Sense of Your Investigation Results*.

Lesson 8: How does the energy transferred from a battery to a wire coil compare to the energy transferred from a computer to a speaker?

Lesson Question

In the last lesson, you considered the energy involved in the magnet systems. Now we will use our explanations as we think about energy transfers through electrical systems and our speaker.

Scientists Circle

1. Meet with your class in a Scientists Circle to discuss what parts of your system (other than the magnets) might be involved with energy transfer.

Turn and talk

2. Discuss the following questions with your partner.
   - What evidence do we have that connecting the wire coil to a battery provides it energy to do something?
   - What are some other systems that need to be connected to a battery in order to do something?

Be prepared to discuss your answers with the class.

Investigating Two Systems

In your notebook

3. We will be investigating two systems--an electrical system with a battery and another system connected to a computer. Set up your notebook following these guidelines:
   - Add a title to a two-page spread of your notebook as shown in the illustration below. The title should be "How does the battery vs. the computer provide energy to the electromagnet?"
   - Make a T-chart that has 2 columns and 4 rows to keep track of each investigation question and the related explanation or solution. Below this leave space to record hypotheses for a 5th investigation.

How does the battery vs. the computer provide energy to the electromagnet?	
Investigation Question	Explanation or solution
Hypotheses:	

With your group

4. Record the first investigation question, *How can we connect a battery to a lightbulb in order to provide it energy?*
    - Conduct the first investigation with your team.
    - Draw and label a diagram of your solution. You can also use words to explain what you did to light the lightbulb.

With your class

5. Share your results with your class as your teacher directs.

An *electric circuit* is a complete, unbroken path for electric current to flow. *Electric current* transfers energy in a circuit.

In your notebook

6. Add labels and drawings to the sketch you made of the circuit for this investigation to indicate how the current flows through your circuit.

Predictions

Turn and talk

7. You have thought about how energy is transferred from the battery to the lightbulb through an electric current.
    - Do you think the music player is transferring energy to a speaker through an electric current when it is producing sounds?
    - How could connecting the lightbulb to the headphone output jack (AUX) wires from a computer help us figure that out?
    - Be prepared to share your ideas with the class.

Investigating Volume Effects on a Speaker and a Lightbulb

In your notebook

8. Add the question "Is the computer providing electric current to the speaker to produce sounds?" to your T-chart.

With your class

9. Watch as your teacher connects a speaker to the computer that is running the sound generator.
   - Record your explanation or solution in your T-chart.
   - Participate in a class discussion of the demonstration.

In your notebook

10. Add the question "How can we change the amount of energy transferred by an electric current to a lightbulb"? to your T-chart.

Exploring How to Get the Lightbulb to Shine Brighter

With your group

11. Use the materials provided to conduct your investigation to answer your new question.
    - Give different group members a chance to engage with these materials.
    - Take 4 minutes to conduct your investigation. When the timer goes off, push the materials to the side and complete the appropriate row on your T-chart.

Making Predictions of Frequency Effects on the Lightbulb

With your class

12. Share with your class what you discovered in your most recent investigation.

On your own

13. Answer the question "What is changing in the electronic music player system when you increase the volume of the sound being produced?"
    - Turn your exit ticket over. You will answer more questions later.

With your class

14. Think back to the *Sound Unit* and the app you used in the last investigation. This app is from the *Sound Unit*, and we used it in our last investigation to change the volume of sound. What other property of sound could you change with the app?

On your own

15. Return to your exit ticket and respond to questions B-D.
    - Once you are finished, turn your exit ticket in.
    - Be prepared to discuss your ideas with the class.

With your class

16. Discuss the following questions with your class.
    - What did we figure out about energy in the system when the speaker is producing a quieter vs. a louder sound?
    - What new question(s) did that raise for us at the end of last class?
    - What investigation did we think it would make sense to do next with the speaker and lightbulb to answer this question?

In your notebook

17. Add the fourth question to your T-chart: "How does the electric current compare for different frequency sounds the computer tries to produce?"

Investigating Changing Frequency with a Light Bulb

With your class

18. Participate in a whole-group interactive demonstration.
    - How do you think what you saw happening at low frequencies is related to the amount of energy being transferred by the electric current?
    - What new questions do you have?

Investigating Changing Frequency with a Coil and Compass

With your class

19. Participate in a whole-group interactive demonstration with the sound generator and the coil and compass.
    - Record your explanation in the second column of your T-chart. Use your observations how both the compass and light bulb were affected by changes in frequency for this explanation.

Turn and talk

20. Consider the results of the demonstration. Which of the following (a or b) do you think is happening? Why?
    - The current is repeatedly switching from on to off (at a certain frequency.
    - The current is repeatedly doing something else that causes the needle and speaker to move back and forth (at a certain frequency).

21. Be ready to share your ideas with the class.

In your notebook

22. There are two possible causes to what you observed. Write the two different causes (a and b) in your notebook. Add additional notes to describe what you have figured out.

Exploring How to Get the LED to Light Up

With your group

23. Using the LED bulb, conduct an investigation to figure out how to provide this new bulb energy to light up.
    - Record your results in your notebooks.

Forming Hypotheses

With your class

24. Participate in a class discussion summarizing what you discovered about the LED light bulbs.

In your notebook

25. Review your responses in #22. Add the idea that something could be flipping in part of the system. Do this by updating B so that you replace the phrase "doing something else" with the phrase "flipping something in the system".

With your group

26. Turn the two possibilities from the previous task into two competing hypotheses about what is changing in the electric current when the computer produces a different-pitched sound. Explain how we could use the LED to test both. Work with your group to discuss and record these two alternate hypotheses in your T-chart.

Investigation of Changing Frequency with the LED

With your class

27. Participate in a whole-group interactive demonstration using the LED and changing the frequency from the sound generator.
    - Use the two demonstrations to evaluate your two competing hypotheses.
    - On your T-chart, circle the hypothesis that is supported by evidence from the demonstration.
    - Strike through the hypothesis that is refuted by the evidence.

Turn and talk

28. Discuss the following with a partner. Be ready to share your ideas with the class.
    - Which hypothesis about the LED light was supported?
    - Which hypothesis about the LED light was refuted?
    - What is your evidence?

In your notebook

29. Make another T-chart in your notebook. Title the first column "What we figured out". Title the second column "Questions we have".
    - Work with your class to add to the first column of your chart.

On your own

30. Add new questions you have to the second column of your chart.
    - Draw a line under your questions.
    - Be prepared to share your ideas with your class.
    - Add new questions from your classmates below the line that you drew.

Reading: Music to My Ears

With a partner

31. Investigate the answers to your questions by reading *Music to My Ears* with a partner. Take these steps as you read:
    - Pause at each bolded question. Wait for your partner and then discuss your answer to the question.
    - Place a check mark by statements in the reading that match what you recorded in the first column of your T-chart.
    - Place a star by statements that help answer your new questions.
    - When both of you are finished, quietly talk about and try to answer the new questions you recorded in the second column of your T-chart.

On your own

32. Think about what you learned in the reading. What can you add to the first column of your T-chart? Record new science ideas you have figured out in this column.
    - Be prepared to share your ideas with your class.

Lesson 9: How do the magnet and the electromagnet work together to move the speaker?

Navigation: What have we been up to?

With your class

1. What are we trying to figure out about the speaker?

2. What are some of the surprising new things we have figured out in Lessons 7 and 8?

Tracing Cause-and-Effect Relationships

In your notebook

3. Look back in your notebook for some additional ideas about cause-and-effect relationships we have uncovered in Lessons 7 and 8.

With your group

4. Complete the table with changes we made to the system and effects we observed when we made those changes.

Turn and talk

5. Which of these cause-and-effect relationships do we need to explain how the speaker makes sound?

Make Models in Groups

With your group

6. Refine a model of the system that captures what we've figured out to help explain the speaker.
   - Be sure to include all the important parts of the system.
   - You may need to draw the system multiple times to represent all the important relationships.
   - Record your own version of the model in your notebook for your teacher to provide feedback on.

Navigation and Home Learning

On your own

7. Where else do we see electromagnets being used to make things move? Do one (or more) of the following to answer this question:
   - Ask your **family and community members** if they know of something that uses electromagnets to make things move.
   - With a friend or family member, use a computer or phone to **search the internet** for electromagnets that make things move.
   - Check the library for **books or articles** about electromagnets that make things move. Ask a librarian for support!

Build a Consensus Model in a Scientists Circle

Scientists Circle

8. Discuss the following with your class:
   - What ideas should we include in the model?
   - How should we represent those ideas?
   - Do we have evidence to support them?

9. Add the consensus model to your Progress Tracker.

10. What questions from the DQB can we answer now?
    - If you see a question we have answered, record an answer on the same card.
    - Move the answered question to the "Questions we have answered" area.

Broaden to Related Phenomena

Home Learning

11. Recall our home learning question: *Where else do we see magnets being used to make things move?*
    - Share what you figured out with a partner.
    - Be prepared to share with the Scientists Circle.

With your group

12. Use the close reading protocol to make sense of your assigned reading.

Before

a. Record the name of your group's reading at the top of your group's poster.

b. **With your group:** Identify the question(s) you are trying to answer. Record them in your notebook for later.

During

c. Read once **individually** for understanding to see what the reading is about.

d. Read a second time out loud **with your group** to highlight a few key ideas.

After

e. Summarize the key ideas in your notebook. Make sure to include

- a description and/or diagram of how the device works that your group read about,
- why it is important that the device relies on an electromagnet, and
- at least one key similarity and one key difference between your device and the speaker that your group noticed.

With your group

13. Follow your teacher's directions and move into new groups to share your learning using the following procedures:

Small-Group Talking Stick - Round 1

- Pass around a pencil as a "talking stick" to take turns having each person share their device and what it does (one minute per person).
- As each person shares, think about (1) an important similarity between your devices and (2) an important difference between your devices, but don't add your own ideas yet.

Small-Group Talking Stick - Round 2

- Pass the talking stick around again to have each person share one similarity they noticed among all three devices and one difference they noticed. Take one to two minutes per person.

Small-Group Open Discussion - Round 3

- With your group, brainstorm what we would need to change about the electromagnets in our classroom for them to work in this new application, and why.

Revise and Add Questions to the DQB

With your class

14. What would we need to do to make an electromagnet strong enough to move cars and trains?

15. What questions do we see on the DQB about what changes might affect the strength of forces in a magnetic field?

With your group

16. Revise an existing question about changing the strength of magnetic forces or come up with a brand-new one. Use cause-effect language to make it clear **what relationship we need to investigate** in order to answer the question:

When we_[change to system]_will we observe_[effect on system]_?

Lesson 10: How does distance affect the strength of force pairs in a magnetic field?

Navigation

Turn and talk

1. What are some factors that might affect the strength of magnetic forces?

Co-designing an Investigation

Turn and talk

2. Develop cause-effect relationships that you predict will tell us something about the effect of distance on magnetic forces. Be sure to use the sentence frame we have been using to write hypotheses:
If __*explanation or theory*__, then when we __*cause*__ we will observe __*effect*__.

With your class

3. Work with your class to identify investigation variables that correspond to the cause-effect relationship in the hypothesis.
   - What is the part of the system we are varying or changing? *This is called our independent variable, and it is the cause in our cause-effect relationship.*
   - What is the part of the system we think we will see change as a result? *This is called our dependent variable, and it is the effect in our cause-effect relationship.*
   - What do we need to keep exactly the same? *These are controlled variables.*

Navigation

Turn and talk

4. How can we organize and analyze data we gather in this investigation? Be ready to share your ideas with the class next time.

Navigation

With your class

5. What were your ideas for how we can organize and analyze data we gather in this investigation?

On your own

6. In your notebook, make a rough sketch of what you would expect the graphs to look like if
   - the data support our hypothesis.
   - the data do not support our hypothesis.

Be ready to share with your class.

Conduct an Investigation and Graph Results

With your group

7. Follow these procedures to conduct the investigation:
    1. Tape one disc magnet to the scale, sticker side up.
    2. Change the units on your scale to ounces (oz) and tare your scale.
    3. Have one group member hold the ruler on the table behind the scale, NOT ON THE SCALE - THIS WILL AFFECT YOUR MEASUREMENTS. Get eye-level with the top of the magnet taped to the scale. Record the number from your ruler that is in line with the top of the magnet. This is your starting reference to help you figure out when the magnets are 1-6 cm apart.
    4. Have another group member slowly move the 4 stacked magnets closer to and farther from the magnet taped to the scale. Make observations of the change in force on the scale as the set of 4 magnets moves closer to and farther from the magnet on the scale.
    5. Use your starting reference number on the ruler to move the magnets from 1 cm apart to 6 cm apart and measure the force on the scale. Be careful to keep the magnets directly in line as you take the measurements.
    6. Record these changes in force in your data table to the nearest hundredth of an ounce (0.01). Flip the set of four magnets and repeat steps 1-6.

With your group

8. Discuss these questions with your group and be ready to share your thinking with the class:
    - When we changed the distance between magnets, what pattern in attractive forces did you observe? Use descriptions from math class where appropriate.
    - When we changed the distance between magnets, what pattern in repulsive forces did you observe? Use descriptions from math class where appropriate.

Make Sense of the Relationship between Distance and Magnetic Force

In your notebook

9. Use the I^2 strategy to make sense of your graph.
    - Step 1: Identify - What I see
        - Identify any changes, trends, or differences you see in the graph or figure.
        - Draw arrows and write a "What I see" comment for each arrow.
        - Be concise in your comments.
        - Do not try to explain the meaning at this point.
    - Step 2: Interpret - What it means
        - Interpret the meaning of each "What I see" comment by writing a "What it means" comment.
        - Do not try to interpret the whole graph.
    - Step 3: Caption
        - Write a caption for the graph.
        - Start with a topic sentence that describes what the graph shows.
        - Then join each "What I see" comment with its "What it means" comment to make a sentence.
        - Build a coherent paragraph out of your sentences.

Building Understandings

Turn and talk

10. When we changed the independent variable, what pattern in the dependent variable did we observe?

With your class

11. Discuss these questions:
    - When we changed the distance between magnets, what pattern in attractive forces did you observe? Use descriptions from math class where appropriate.
    - When we changed the distance between magnets, what pattern in repulsive forces did you observe? Use descriptions from math class where appropriate.

On your own

12. Write an explanation, supported by evidence from the investigation, to answer the lesson question: *How does the distance between magnets affect the strength of forces on the magnets?*

In your explanation, include the following:

- **Claim:** *Explanation or theory* from our hypothesis
- **Evidence:** When we changed the *independent variable* we observed the following pattern in the *dependent variable*.
- **Reasoning:** Explain why the evidence supports the claim.

Lesson 11: What else determines the strength of the force pairs between two magnets in a magnetic field?

Navigation

In your notebook

1. Write the lesson question in your notebook:
   What else determines the strength of the force pairs between two magnets in a magnetic field?

Make Predictions

With your class

2. Discuss the following questions with your class:
   - What do we think we can do to the system to affect the strength of magnetic forces between two magnets?
   - What about between a magnet and an electrified coil?

Designing an Investigation

With your group

3. Work with your group to design an investigation to determine the effect on magnetic forces of your assigned factor.
   - Hypothesis: Develop a cause-effect relationship you predict will tell us something about the effect of your assigned factor on magnetic forces.
   - Variables: Work with your group to determine the variables in your investigation.
   - Investigation Plan: Plan the investigation using the rubric provided by your teacher for guidance.

4. Work with your group to determine how you will record and analyze your data.
   - Develop a data table that reflects the number of measurements you will take and appropriate units of measure.
   - Determine labels for the *x*- and *y*-axes of your graph.
   - Be sure to consider how you will record both attractive and repulsive forces in your table and on your graph.

5. On part E of your handout, use the graphs provided to make predictions about what you would expect a graph of your results to look like if
   - the data support your hypothesis and
   - the data <u>do not</u> support your hypothesis.

Peer Feedback on Investigation Plans

With your group

6. Follow your teacher's directions to give feedback on the investigation plan of another group.

Conduct an Investigation and Graph Results

With your group

7. Conduct your investigation.
    - Tape a magnet to the scale.
    - Press the units button on the scale until ounces (oz) is showing.
    - Tare the scale with the magnet taped to it when there is no other magnet or metal object nearby.
    - Record this measurement in your notebook. Tare the scale again between measurements, when there is no other magnet or metal object nearby, if the number changes.
    - Carry out your observations and fill out your table. Remember: you will only need to change the independent variable that your group is investigating. All other variables should be the same for each value of the independent variable you test.

Make Sense of Data

With your group

8. Determine appropriate scales and labels for the *x*- and *y*-axes of your graph.
9. Graph your data.

Interpret Results

In your notebook

10. Use the I^2 strategy to make sense of your graph.
    - Step 1: Identify - What I see
        - Identify any changes, trends, or differences you see in the graph or figure.
        - Draw arrows and write a "What I see" comment for each arrow.
        - Be concise in your comments.
        - Do not try to explain the meaning at this point.
    - Step 2: Interpret - What it means
        - Interpret the meaning of each "What I see" comment by writing a "What it means" comment.
        - Do not try to interpret the whole graph.

- Step 3: Caption
    - Write a caption for the graph.
    - Start with a topic sentence that describes what the graph shows.
    - Then join each "What I see" comment with its "What it means" comment to make a sentence.
    - Build a coherent paragraph out of your sentences.

Make a Claim

With your group

11. With your group, create a poster that includes the following elements:
    - Investigation hypothesis
    - Do your data support or refute your hypothesis?
        - If your evidence does not support your hypothesis, describe additional data you want to collect.
    - A visual representation of the relationship you described in your claim

Gallery Walk

With your group

12. Follow your teacher's instructions for a gallery walk.
    - Move from poster to poster according to your teacher's instructions.
    - At each poster, identify an interesting or confusing pattern you notice in the data.
    - Record that pattern and leave feedback on the poster.

Building Consensus

With your class

13. Meet in the Scientists Circle and share investigation findings with your class.

Turn and talk

14. We have been measuring individual forces in one place near the magnet. What does this tell us about the magnetic field around a magnet?
15. What could we do to figure out how the changes we made are actually impacting the whole magnetic field?

Computer Interactive

With your group

16. What predictions can you make about what the magnetic field will look like when we change the variables in the interactive? Record your predictions as diagrams on your handout.

17. Use the interactive to replicate all the investigations we did in class. Record your results on your handout.

Home Learning

On your own

18. Fill out the cause-effect chain diagram on *Cause-Effect Chain* as best you can.
    - You can modify the diagram to add more boxes to the chain of cause and effect if needed.

Lesson 12: What cause-effect relationships explain how magnetic forces at a distance make things work?

Navigation

In your notebook

1. Write the lesson question in your notebook:

   What cause-effect relationships explain how magnetic forces at a distance make things work?

Reviewing Cause-Effect Relationships

With your class

2. Compare your cause-effect chain with a partner and then with your class.

Apply Our Ideas

Scientists Circle

3. How have our investigations in Lessons 10 and 11 helped us understand how these very big electromagnets work?
4. How have our investigations in lLessons 10 and 11 helped us understand the speaker better?
5. What can we say now about the relationship between forces, energy, and magnetic fields?

Revisit the Driving Question Board

With your class

6. Follow your teacher's directions and use sticker dots to indicate
   - questions we agree that we can answer,
   - questions that we have at least a partial answer to, and
   - questions we cannot answer at all.

Reflection

In your notebook

7. Find a new page in your notebook and title it "Reflection". Under the heading, record your answers to these questions:
   - What was most challenging in this unit?
   - What was most rewarding?

Demonstrate Understanding on an Assessment

Turn and talk

8. Choose one of the applications that we have learned about: maglev trains, electric motors, junkyard magnets, or speakers. Discuss the following question in relation to the application you chose: *Why is it useful to have an electromagnet instead of just using a permanent magnet(s) in the system?*

In your notebook

9. If you could use an electromagnet to apply a force to something or make something move, what would you design? Be ready to share your ideas with the class.

References

Composition of Metal Objects

Object	What the metal is made of					
	Iron (Fe)	Carbon (C)	Nickel (Ni)	Zinc (Zn)	Copper (Cu)	Aluminum (Al)
paper clip	98.0%	2.0%	0.0%	0.0%	0.0%	0.0%
iron nail	99.9%	0.1%	0.0%	0.0%	0.0%	0.0%
quarter	0.0%	0.0%	8.0%	0.0%	92.0%	0.0%
penny (minted after 1981)	0.0%	0.0%	0.0%	97.5%	2.5%	0.0%
compass needle	98.0%	2.0%	0.0%	0.0%	0.0%	0.0%
coil of wire	0.0%	0.0%	0.0%	0.0%	100.0%	0.0%
aluminum foil	0.0%	0.0%	0.0%	0.0%	0.0%	100.0%

Guidance for Student Facilitators

Guidance for Student Facilitators

As your peers share ideas, record the ideas that we agree on with the help of your teacher.

1. Before you start, remind your peers about our classroom norms.

Say: *Think of a norm that you would like to work on. Share with a partner the norm you chose.*

2. Remind your peers how to communicate respectfully.

Say: *It's OK to share an idea that you are not sure about. It is also OK to disagree with somebody's ideas. But please disagree respectfully.*

3. Remind your peers about the purpose of the discussion.

Say: *The purpose of this Consensus Discussion is to build a model that we all agree on. The model will explain how magnetic forces cause vibration in the speaker. We will use ideas we have figured out in Lessons 1 through 8.*

4. Facilitate consensus building. Use the prompts below to help your peers come to consensus. Feel free to reword these questions or add your own questions.

Ask about what should go in the model:

- *What new parts do we need to include?*
- *What parts from the old model do we not need anymore?*
- *How should we represent both cause and effect for each of these relationships?*
- *What additional versions of the system do we need to include in order to capture each cause-effect relationship?*

Ask your peers to support or challenge proposals:

- *What ideas are we in agreement about?*
- *Are there still places where we disagree? Can we clarify these?*
- *Are there still areas of controversy, confusion, or discontent?*
- *Who feels like your idea is not quite represented here?*

Help your peers come to consensus with conflicting ideas:

- *How are these explanations similar? How are they different?*
- *How could we modify what we have so that we account for the evidence we agree is important to consider?*
- *Is there more evidence or clarification needed before we can come to consensus? What is that?*

If you need a break, call a time-out. A *time-out* is a time when the rest of the class waits so that you and your co-facilitators can confer.

The Identify and Interpret (I^2) Strategy - Student Guide

Have you ever looked at a graph or figure and felt overwhelmed by it? Often there is a lot of information on graphs and in figures. The Identify and Interpret (I^2) strategy helps you make sense of graphs, figures, sketches, data tables, and other ways to represent data. This strategy helps you break down the information into smaller parts. To do this, you first identify what you see in the graph or figure. Then you interpret each of those observations by deciding what they mean. Once you have determined what the smaller parts of the graph or figure mean, you are ready to put all the information together. To do this, you write a caption. You have probably seen captions under figures in textbooks or magazines. Captions are a summary of the information in the graph or figure. They are written in complete sentences. Captions help you show your understanding of the material you are studying. To help you understand how to use the I^2 strategy, look at the following example. This example will help you make sense of a graph. This graph shows the average monthly temperatures in one US city.

I^2 step	Example
Step 1: Identify ("What I see" comments) ▪ Identify any changes, trends, or differences you see in the graph or figure. ▪ Draw arrows and write a "What I see" comment for each arrow. ▪ Be concise in your comments. These should be just what you can observe. ▪ Do not try to explain the meaning at this point.	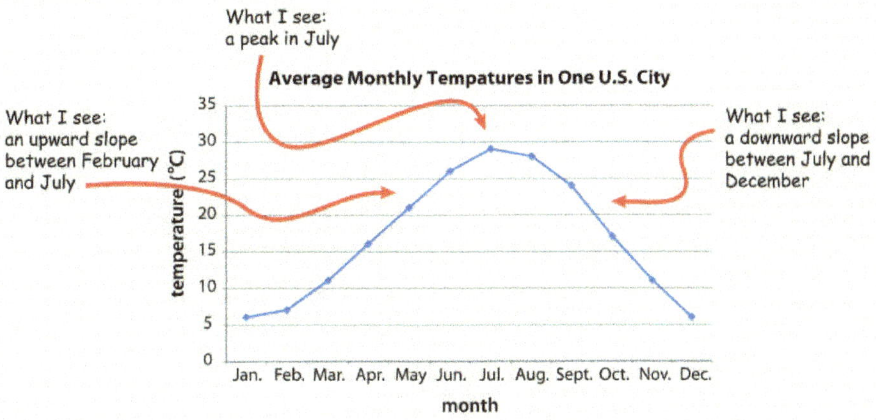 For this example, there are arrows drawn that point to the two trends and the change. Notice that the arrows point to the general upward and downward trends, not to each data point. A "What I see" comment describes what each arrow points to on the graph.

I² step	Example

Step 2: Interpret ("What it means" comments)

- Interpret the meaning of each "What I see" comment by writing a "What it means" comment.
- Do not try to interpret the whole graph or figure.

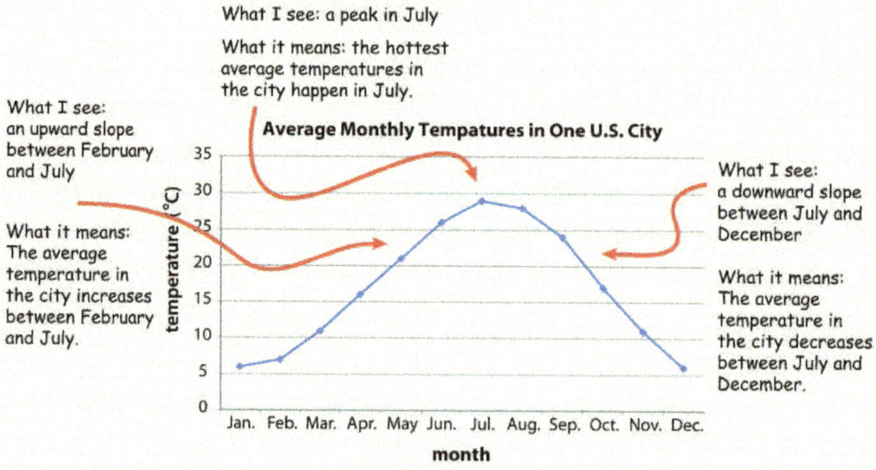

In this example, "What it means" comments were added to each "What I see" comment. The "What it means" comments explain the changes, trends, and differences that were identified in Step 1.

I² step	Example

Step 3: Caption

- Write a caption for the graph or figure.
- Start with a topic sentence that describes what the graph or figure shows.
- Then join each "What I see" comment with its "What it means" comment to make a sentence.
- Build a coherent paragraph out of your sentences.

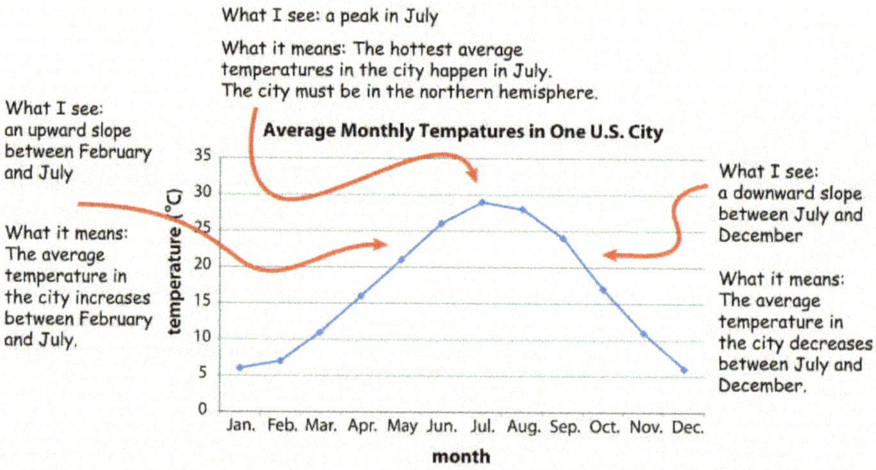

This graph shows the average temperature in a city over a year. There is an upward slope from February to July, showing that there is an increase in the average temperature during these months. There is a downward slope from July to December, which means the average temperature decreases during this time. There is a peak in July, which shows that the hottest average temperature in the city happens in July. The peak in July also means that the city must be in the Northern Hemisphere.

In this example, the first sentence of the caption describes what the graph shows. Then each "What I see" comment was combined with its "What it means" comment to form complete sentences. Those sentences make up a paragraph that describes each part of the graph.

Readings

Finding the Way

Close Reading Strategies

As you read, use the following close reading strategies to support your understanding.

1. Identify the question(s) you are trying to answer with the reading.
2. Read once for understanding to see what the reading is about.
3. Read a second time to highlight a few key ideas that help answer the questions you had.
4. Summarize the key ideas in your own words, in diagrams, or both.
5. Jot down new questions that the reading raises for you.

Finding the Way

Have you ever been lost? What did you do? Today, many people use their smartphones to navigate in unfamiliar places. But what would you do if you didn't have a smartphone or a computer to help you find your way?

Ancient civilizations used other methods to figure out where they were and which way they needed to go. Familiar landmarks, the flight paths of birds, weather patterns, and the direction and location of the Sun and stars helped travelers move from place to place. And ancient travelers also relied on compasses to help them figure out what direction they were facing.

Chinese scientists made the first known magnetic compasses over 2000 years ago. They used a naturally occurring magnetic rock called *lodestone,* pictured here. Early magnetic compasses used a magnetized needle attached to a lightweight piece of wood or cork. The wood or cork floated in a small water dish. Floating in the water allowed the needle to turn to show direction. Have you ever used a magnetic compass, or seen someone else use one?

A magnetic compass can be used by hikers to navigate. Modern magnetic compasses typically have a magnetized needle mounted over a display that shows the four cardinal directions–north, south, east, and west. The needle of a magnetic compass always points north no matter which way it is positioned. Why do you think this is?

The Earth has a magnetic field, which is why the compass needle always points toward the North Pole. In class you used magnetic compasses to map the magnetic field of a bar magnet. You also used a computer simulation to better visualize this field. You probably noticed that the needle of the magnetic compass always points towards the <u>south</u> pole of the magnet. So why does a compass needle always point to the <u>North</u> Pole of Earth?

To understand this, we need to differentiate between the *geographic poles* of Earth and the *magnetic poles* of Earth. The geographic poles of Earth refer to the northernmost point and the southernmost point on the planet. We call these the North and South Poles. They are aligned with the axis of the Earth. These are the points around which the entire planet spins over the course of a day, making one full rotation every 24 hours.

But the geographic North Pole is not the same as the magnetic north pole of the Earth. Where do you think the magnetic north pole of the Earth is located?

The geographic North Pole and the magnetic north pole are at opposite ends of the Earth!
The North Pole of the Earth is actually very close to the magnetic <u>south</u> pole. This should sound familiar - you saw that a compass points toward the south pole of a magnet in this lesson. You might hear people say that a compass points toward the North Pole. Now you know that this is only true of the geographic North Pole - it is actually pointing toward the magnetic south pole! In addition, the magnetic field does not line up perfectly with the axis of the Earth. This is important for anyone navigating with a compass. The distance between the geographic North Pole and the magnetic south pole affects the accuracy of a magnetic compass. Navigators must make adjustments to be accurate. The effect is more significant the closer one gets to the geographic poles and is less noticeable near the equator.

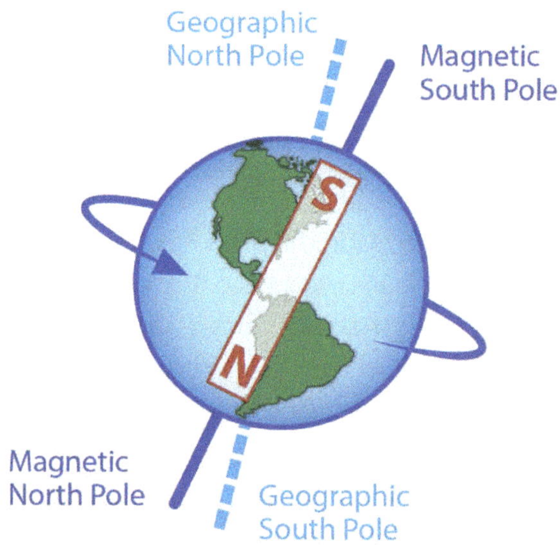

Look for directions for making your own compass here: https://media.nationalgeographic.org/assets/file/MakeyourownCompass.pdf

Music to My Ears

Instructions for reading

1. Read silently until you get to a question in bold.
   a. Pause and wait for your partner.
   b. Quietly discuss your ideas to answer the question.
   c. Continue reading until the next bold question.
2. As you read about them, place a check mark by things we figured out from our two-column chart.
3. As you read about new ideas to answer our questions, place a star by them in the reading.
4. When you complete the reading, discuss the questions from your two-column chart with your partner.
   a. How would you answer the questions now?

Music to My Ears

Think about people you have seen use headphones to listen to music on a phone or to complete an assignment on a computer in school.

We have learned a lot about how speakers work.

1. What do you think headphones have in common with the speakers you investigated in science class so far?

If you dissect headphones, this is what you would see:

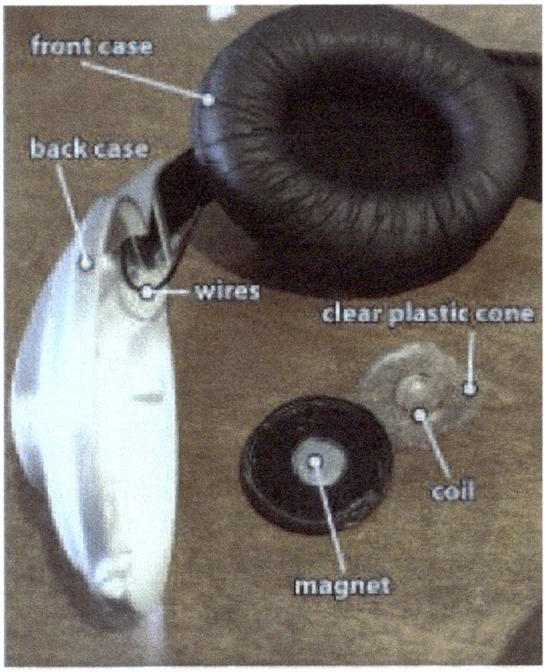

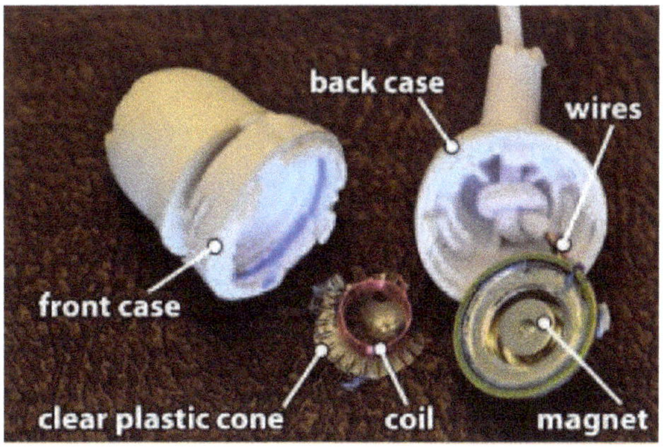

You may have noticed in the photos above that headphones contain many of the same parts as a speaker (including a magnet and a coil of wire). Those parts are small compared to the parts of the speakers you made in class.

We learned that attractive and repulsive forces between the magnet and the coil lead to vibrations in the speaker. These vibrations produce the sound we hear. The attractive and repulsive forces inside a speaker happen when an electric current flows through the coil. The current causes the coil to behave like a magnet. We call the magnetized coil an *electromagnet*. The permanent magnet in a speaker is usually fixed in place, but the electromagnet can move back and forth.

Like all speakers, headphones must be connected to a source of electric current in order to work. Electric current flows through a speaker system following the path of a circuit. A *circuit* is an unbroken path for electric current to follow.

2. What about wireless headphones? How do they work if they are not attached to a source of electric current?

Wireless headphones allow you to walk away from your phone, computer, or music player, and still hear the sound. These headphones still work in the same way as the speaker. There is a tiny magnet inside, and a tiny electromagnet. But instead of a wire connected to a computer, there is a battery. Electronics inside the headphone change the current coming from the battery in order to change the forces in the electromagnet.

Instead of sending the changing electric current directly down a wire to the speaker, the computer needs to send a message to the headphones, that tells the headphones how to change the current that is coming from the battery. The message that a computer sends to headphones that are plugged in is a continuous signal that will be directly translated to the speaker, and into sound. But in order to send signals through the air, we need to send the signal as a set of instructions, that the headphones can then use to put the signal back together. That set of instructions is in the form of pulses of radio waves known as bluetooth.

3. How does the same electric current power tiny headphones, and huge speakers?

Powerful speakers need more energy, so they will need more electric current. To get enough current, large speakers need to be plugged into amplifiers. An *amplifier* is a device that can increase the electric current. More electrical current creates stronger forces. Stronger forces will transfer more energy, which means the electromagnet will move more and the speaker will produce louder sounds.

You increased the electric current in a circuit when you used two batteries instead of one to light a lightbulb. This provides the circuit with more energy. You saw evidence of more energy when you saw a brighter light. In a speaker system, the computer does something similar to produce louder sounds. The computer increases the electric current in the system to produce louder sounds.

4. Why would producing louder sounds require more energy?

You also investigated how to change the forces in the system from repulsive to attractive forces and back again. One way you did this was to flip the coil around. Another way was to change the orientation of the battery in the circuit so that the ends of the coil are connected to the opposite terminal.

Look carefully at the diagram above.

5. How does this diagram help explain what is causing the forces to change back and forth from repulsive to attractive?

The diagram above shows how current changes in the system when you change the poles that the coil is connected to on the battery. Use your finger to trace the direction of the current in each circuit. Notice that the direction that the electric current flows changes. The poles of the electromagnet change when the direction of current changes. Changing the poles of the electromagnet switches the direction of the forces in the field. Changing the direction of the forces in the field moves the coil in different directions. This is how the computers produce different vibrations and different sounds.

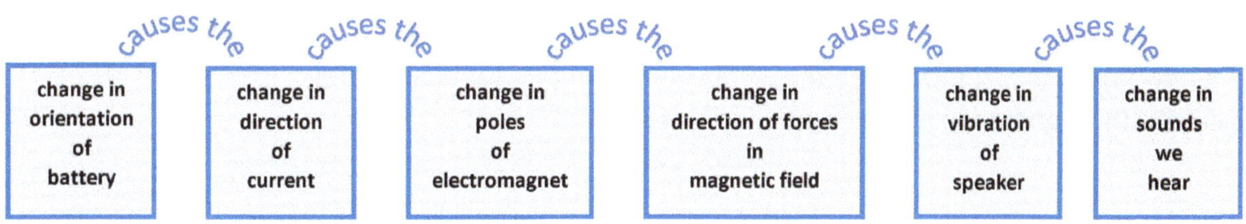

The behavior of the two-color LED was evidence of a changing current. Two-color LED bulbs have two different colored bulbs that light up depending on the direction of current flow. If the current is flowing one direction, you get a red light. If the current is flowing the opposite direction, you get a green light.

You saw evidence of the current changing direction back and forth for very low frequency sound. We could see evidence of the changing direction because we could see the light change from green to red and back again. But our ears cannot detect frequencies this low. When we increased the frequency so that we could hear the sound, our eyes saw only orange light from the LED. That is because the light is switching from green to red so fast that our eyes cannot detect the difference--we see it as orange light.

In the *Sound Unit* we learned that the different high and low sounds (the pitch) we hear are caused by changes in frequency. Pitch, measured by frequency, refers to how frequently the current direction is flipped back and forth. For example, to produce a sound like the one you tested in class (a middle C), the sound source must vibrate at 261 HZ, or 261 times back and forth every second. That means the current must also change direction back and forth 261 times a second.

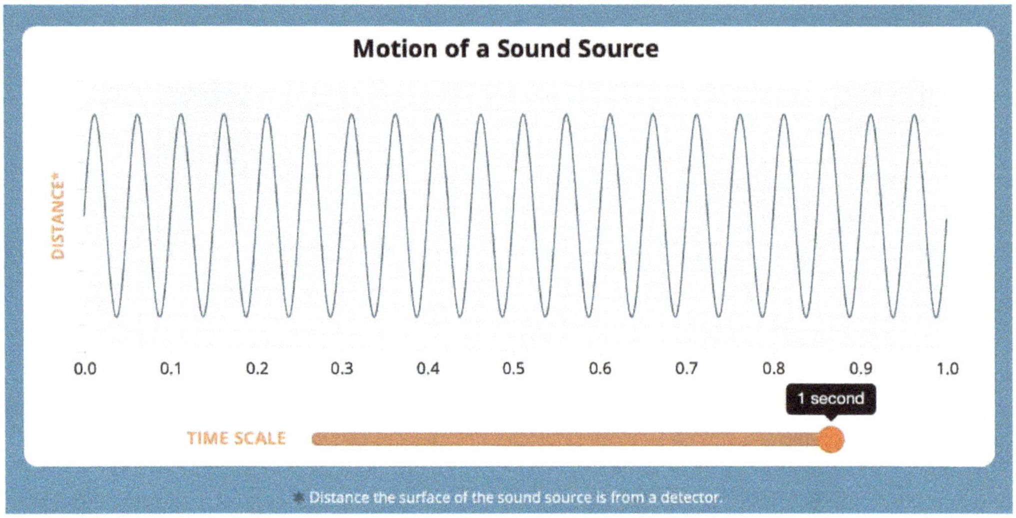

But how does the current change direction so quickly? We know the orientation of the permanent magnet and of the electromagnet inside the speaker doesn't change.

Modern electrical devices, like computers and phones, can switch the direction of current based on directions from a digital music file, like an mp3. Hardware, like the sound card of a computer shown here, can change the direction and the amount of current about 100,000 times a second. The rapidly changing current changes the direction and strength of forces in the system. This changes the way the speaker cone moves, creating different kinds of sound waves. Our ears hear the sound waves as a series of continuous sounds, like the songs you hear through headphones or other speakers.

6. Where does the music come from?

We read that the pattern of current that creates the music we hear is stored in a digital music file, like an mp3. But how did music made by a real person in a studio become a digital music file? When a recording artist sings into a microphone, the signal leaves their mouth as a sound wave, which is a continuous moving pattern of pressure in the air. This kind of signal is known as an analog signal. Analog signals are made up of continuously streaming data.

Consider a clock that has hands moving around in a circle. The hands move smoothly around the face of the clock. This kind of clock is telling you the time in an analog way, constantly telling you exactly what time it is, down to the second. Right now it is showing 10:13 and 10 seconds. But it is in this position only briefly as it moves smoothly to 10:13 and 11 seconds, and then 10:13 and 12 seconds, etc.

A digital clock, on the other hand, is not a continuous signal. It tells you the time in chunks. At 7:10, the digital clock will say 7:10, and it will not change at all until 7:11, when it will flip to display 7:11 very suddenly.

If the signal starts and ends as an analog sound wave, why do we store it digitally on the computer? While it may seem like analog signals are better, because they can tell you information continuously, without interruption, digital signals are much less likely to lose quality as they travel. Digital signals are easier to store on a computer. And digital signals are easier to reproduce. This is why the file stored on your computer is converted into digital.

So where does the music come from? Even though the sound wave leaving the recording artist's mouth is analog, the music producer will use an analog-to-digital converter to change the sound wave into a digital signal stored on a computer. That file eventually makes its way to your computer, where a digital-to-analog converter takes the signal and changes it back into an analog signal in the form of electric current. The current will apply forces to a speaker, send waves of pressure into the air, and fill your ears with music!

What is electric current?

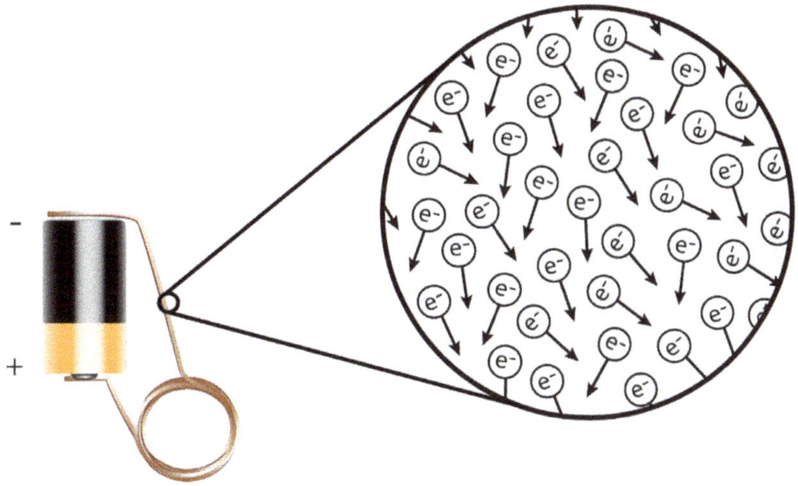

Electric current in a wire is due to the flow of tiny charged particles inside the wire. In the image, you can see these tiny particles. These particles are called *electrons*. Electrons are a part of what makes up the atoms of any object. In metals that conduct electricity well, like copper, these particles can easily be moved around to create an electric current. When the copper wire is connected to the battery, the electrons in the wire start to move all at once in a general direction from the negative to the positive end of the battery. This movement is called a *current*. The arrows in the image above show the direction of the current.

The electrons in a wire are not moving very fast. In fact, we talk about the "drift speed" of electrons, because they are moving so slowly it is like they are drifting through the wire. But when you connect a lightbulb to a battery, the light comes on immediately. Why does this happen if electrons are moving so slow? Electrons do not zoom through the wire in a straight line. Instead, the electrons that are in the wire begin to move all at once and move in a random, zig-zag motion. Since the electrons start to move all at once, energy is transferred immediately when the circuit is connected to the battery.

Go to https://phet.colorado.edu/sims/html/circuit-construction-kit-dc/latest/circuit-construction-kit-dc_en.html to construct your own circuit as part of a computer interactive. Can you turn on the lightbulb without connecting a wire to both ends of the battery? Why or why not?

Magnetic Levitation Trains

The fastest trains in the world. Have you ever taken a train ride? How fast do you think the train was going? Most trains in the United States have a maximum speed of about 100 miles per hour.

But there are types of trains in other parts of the world that travel much faster. One of these types of trains is called a Maglev Train. It has a maximum speed of 270 miles per hour! The Maglev Train between Shanghai Airport (China) and Shanghai City travels a 19-mile distance in 8 minutes. While Maglev Trains move very fast, they are also very smooth and quiet! *What is it about these trains that makes it possible for them to do this?*

What is a Maglev? Maglev Trains get their name from combining the first parts of these two words: magnetic levitation. *Levitation* means an object is suspended, floating, or hovering, without mechanical support. When magnets cause the object to do this, it is called *magnetic levitation*. A regular train has wheels that roll along a track. The Maglev Train has no wheels and does not touch any other material while moving. It is suspended above the track, and so the friction is extremely low. Once the Maglev Train starts moving, there is very little surface friction that would cause it to stop by pushing back on the train.

How does the Maglev Train levitate? We learned that magnets have magnetic fields that surround them. Two magnets can be placed with the same poles facing each other so they repel, or they can be placed with opposite poles facing each other so they attract. In both cases, magnetic fields allow the magnets to exert forces on each other without touching. Think of a magnetic field as the space near a magnet where magnetic force can be detected.

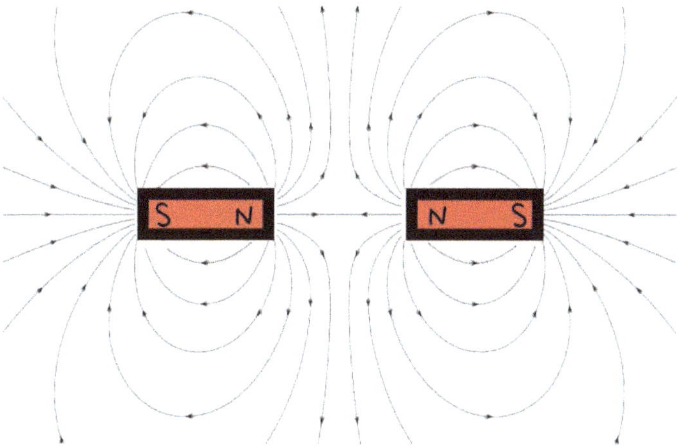

The image shows an illustration of the magnetic field. A diagram of the magnetic field shows us a map of the strength and direction of the forces you would detect if you put a test object near a magnet. In this diagram, the pointers are connected by lines that illustrate the changing direction of the forces on a test object. A test object could be a compass or another magnet, or it could be a piece of metal that is ferromagnetic (easily magnetized temporarily), like steel. Magnets attached to the bottom of a Maglev Train are repelled by the magnetic field of electromagnets attached to the rails below.

What is an electromagnet? Electromagnets levitate the MagLev and propel it forward. See the diagram below. An *electromagnet* is a magnet that relies on electricity to generate a magnetic field. A common way to make an electromagnet is to wrap copper wire into a coil. An electric current must be running through the wire to generate the magnetic field.

How does the Maglev Train move? The Maglev Train is guided forward by electromagnets attached to steel rails. As the train moves forward, the current running through the copper coils on the rails in front of the train switches so that the copper coils have poles opposite to those of the magnets attached to the train. The attractive force will pull the train forward. After the train moves past the copper coils on the rails, the current running through the coils switches again so that the coils are the same poles as the magnets on the train. The repulsive force will continue to push the train forward.

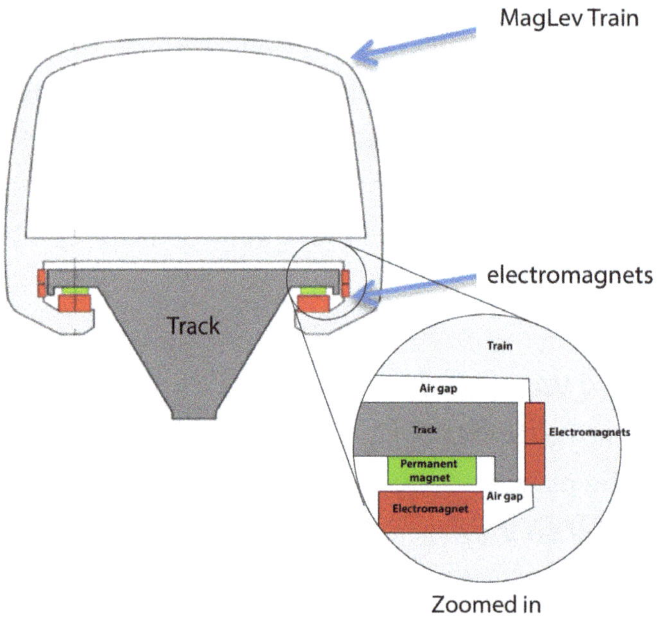

Junkyard Magnets

Jeremy Walker / Science Photo Library

Welcome to the junkyard. Have you ever been to a junkyard? A junkyard is a place where stuff is stored that could be recycled or re-used. For example, old, broken-down cars are something you might find at a junkyard. Metal from old, broken-down cars can be melted down and used to build new things.

Sometimes, the people who work in the junkyard need to move the old cars around. They may need to move the cars to make room for more junk. Or they may need to move the cars onto a truck or a train so they can be broken down somewhere else for scrap metal or parts. *How do you think they move the old cars if they don't drive anymore?*

Junkyard magnets. The workers in the junkyard use enormous magnets to move cars and other large pieces of scrap metal from one place to another.

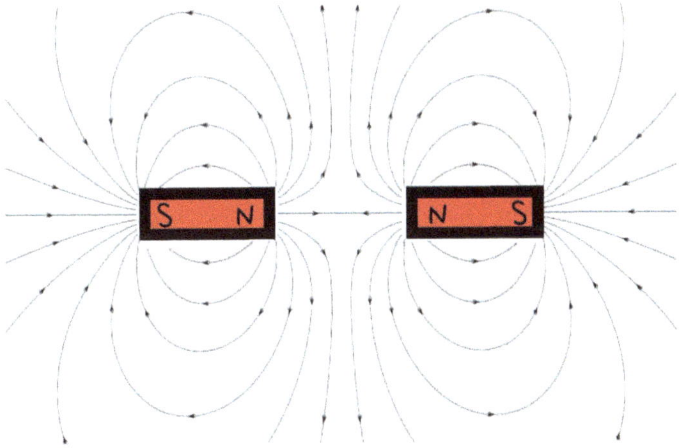

We learned that magnets have magnetic fields that surround them. Think of a magnetic field as the space near a magnet where magnetic force can be detected. The image shows an illustration of the magnetic field. A diagram of the magnetic field shows us a map of the strength and direction of the forces you would detect if you put a test object near a magnet. In this diagram, the pointers are connected by lines that illustrate the changing direction of the forces on a test object. A test object could be a compass or another magnet, or it could be a piece of metal that is ferromagnetic, like steel.

In Lesson 2, we figured out that certain materials are attracted to a magnet, but they cannot be repelled like another magnet can. These materials are *ferromagnetic. Do you remember what kinds of metals are ferromagnetic (attracted to a magnet)? Do you remember what kinds of metals are not ferromagnetic (not attracted to a magnet)?*

A junkyard magnet uses magnetic force to attract ferromagnetic metals found in scrap metal and old cars, like iron. The magnet is usually attached to a crane. The junkyard worker lowers the crane magnet to the car until the car sticks to the magnet. Then the junkyard worker swings the arm of the crane to where the car needs to go. The car is stuck to the magnet, held in place by magnetic force pairs. How do you think the junkyard worker gets the magnet to let go of the car?

What is an electromagnet? The junkyard magnet is an electromagnet. An *electromagnet* is a magnet that relies on electricity to generate a magnetic field. A common way to make an electromagnet is to wrap copper wire into a coil. An electric current must be running through the wire to generate the magnetic field. Unlike a permanent magnet, an electromagnet can be turned on and off easily. That means that when the junkyard worker decides to drop the car, all that worker has to do is turn off the electromagnet and the ferromagnetic metal will no longer be attracted to the magnet. The force pairs between them disappear, and the car is released.

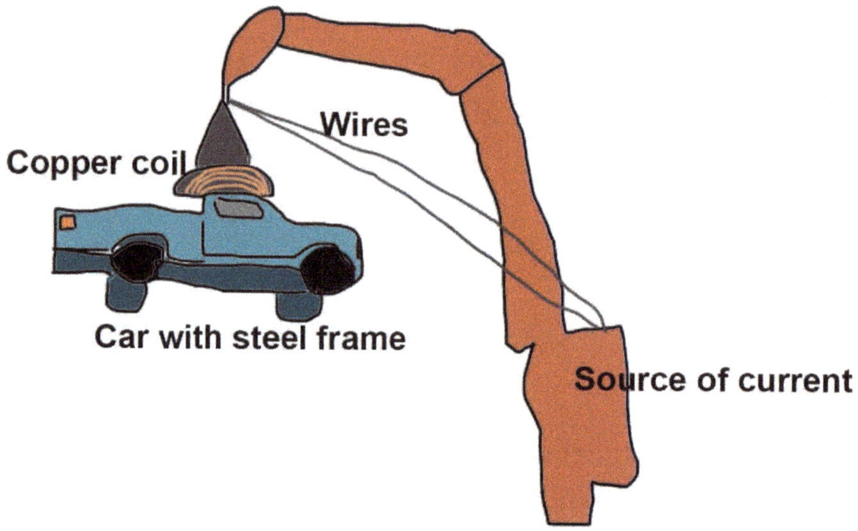

Electric Motors

Electric cars. Electric cars use energy from a battery to move. They are more efficient than gasoline-powered cars, and they produce fewer toxic emissions. How does the electric current from the battery make the wheels of the car turn? The answer is an electric motor.

Electric motor system. The diagram below shows an electric motor. You may recognize some of the parts in the system. There is a coil of copper wire and a battery. A metal rod connects the coil to the gears in the car's transmission system. When the coil spins, it is attached to a metal rod that spins the gears, which work together to spin the wheels of the car.

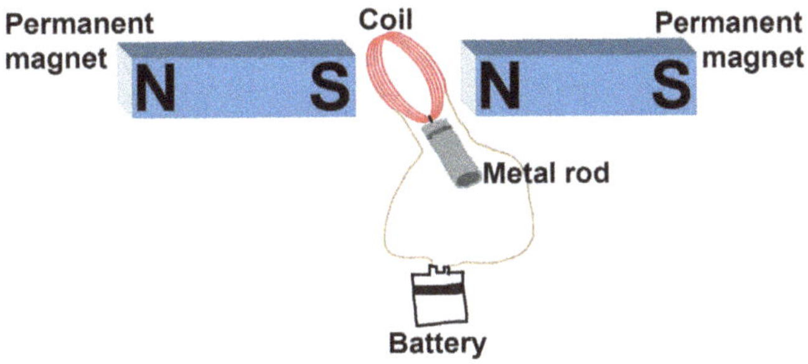

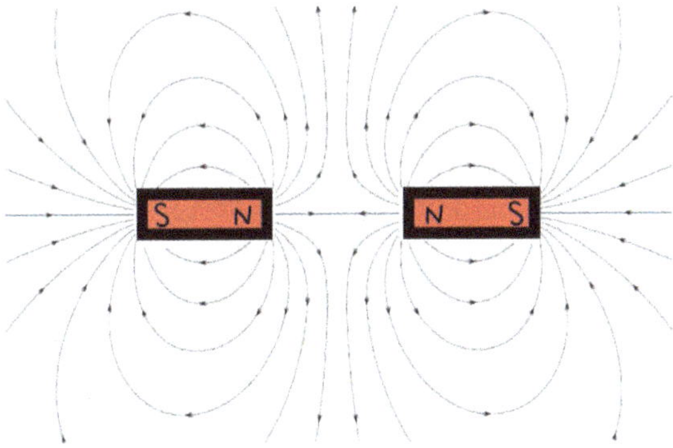

There are also two permanent magnets. One permanent magnet is oriented so that the south pole is facing the coil. The other permanent magnet is oriented so that the north pole is facing the coil. The image at right shows an illustration of the magnetic field between two permanent magnets. A diagram of the magnetic field shows us a map of the strength and direction of the forces you would detect if you put a test object near a magnet. In this diagram, the pointers are connected by lines that illustrate the changing direction of the forces on a test object. A test object could be a compass or another magnet, or it could be a piece of metal that is ferromagnetic, like steel.

Can you figure out how the parts of the electric motor work together to make the coil spin and move the wheels of the car?

The electromagnet. In Lesson 2, we learned that a coil of wire connected to a battery is called an *electromagnet.* An electromagnet relies on electricity to generate a magnetic field. In Lesson 8 we learned that an electric current must be running through the wire to generate the magnetic field. When the electromagnet in a motor turns on, force pairs form between the coil of wire and the two permanent magnets on either side of it. Because like poles repel and opposite poles attract (Lesson 2), the magnetized coil of wire will spin so that its north pole is closer to the permanent south pole and its south pole is closer to the permanent north pole.

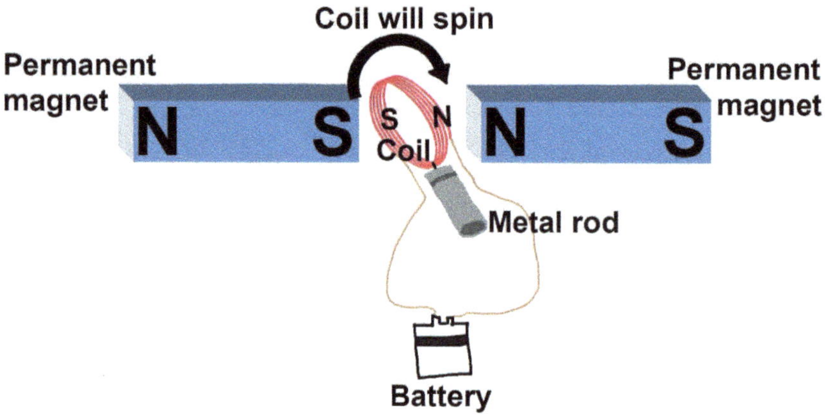

Keeping the motor spinning. In order to keep the motor spinning, the poles of the electromagnet need to switch so that the north pole becomes the south pole. When this happens, the coil will flip again so that the new north pole is facing the permanent south pole and the new south pole is facing the permanent north pole. This happens over and over again so that the coil is constantly spinning.

In Lesson 8, we figured out that to switch the poles of the magnetized coil, we need to switch the direction of the current in the electromagnet. A device embedded in the circuit changes the current direction from clockwise to counterclockwise and back again, over and over. This changes the poles of the magnetized coil back and forth so that the coil will continue to spin.

Strong or Weak?

Scientists believe that at the very center of Earth is a solid core of rock. Around the solid core is another thick layer of rock called the *outer core*. The outer core is between 4,400°C and 6,100°C (7,952°F-11,012°F). Wow, that is hot! The outer core gets hotter than the surface of the Sun! What do you think happens to rock when it gets that hot?

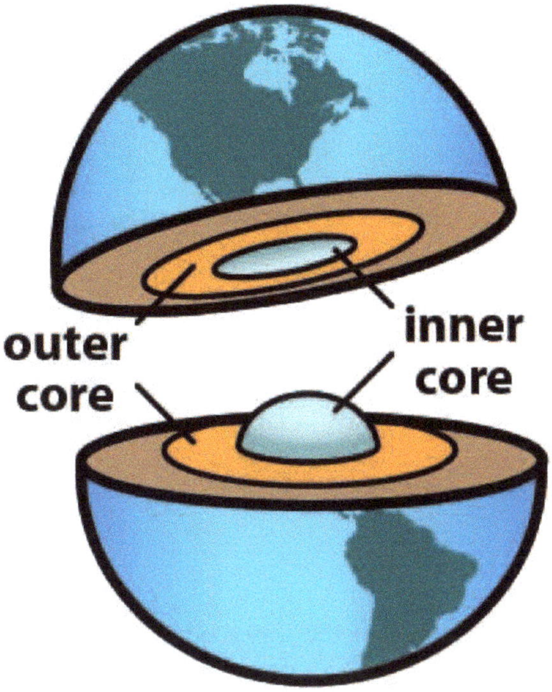

Just like an ice cube melts into liquid water when it gets hot enough, the rock in the outer core melts into liquid rock when it gets hot enough. We call it *molten rock*. This molten rock contains electrically charged particles. Because Earth is rotating, the molten rock in the outer core swirls around in coils that line up with the axis of Earth. Moving, electrically charged particles means that the outer core has an electric current!

What happens when an electric current flows in the shape of a coil? Does that remind you of anything?

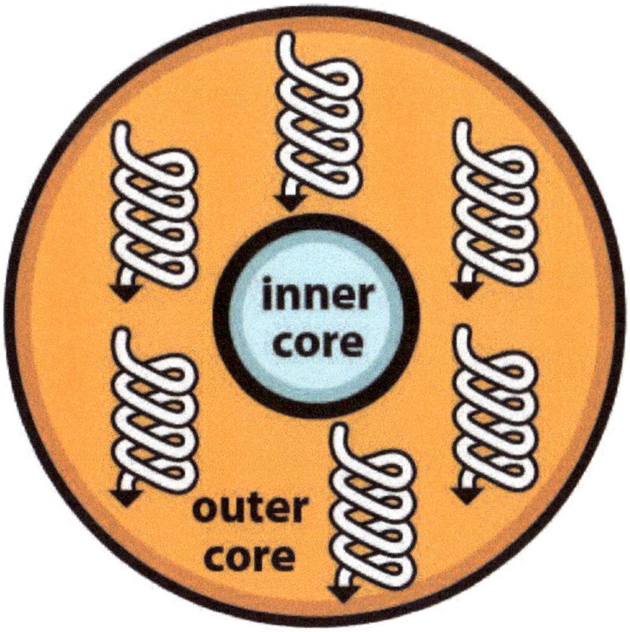

Earth is essentially a giant electromagnet! Earth is really big. Earth is so big that if you could dig a tunnel straight through and pop out on the other side, the tunnel would be about 8,000 miles (almost 13,000 kilometers) long. That is longer than if you drove all the way from the Pacific coast of the United States to the Atlantic coast and then back again!

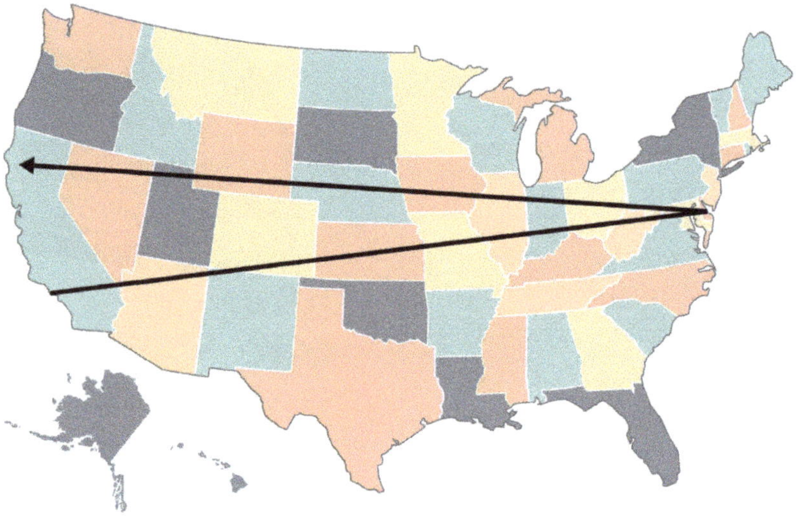

If you were to travel all the way around the middle of the planet and end up back where you started, you would cover more than 24,000 miles.

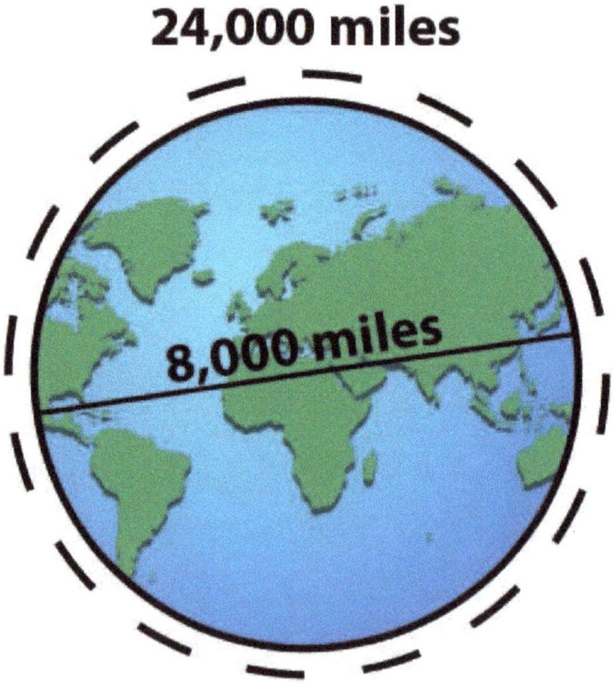

We learned in class that a bigger magnet produces stronger forces than a smaller magnet (if both magnets are made of the same material). Do you think Earth is a weak electromagnet or a strong electromagnet?

Earth's magnetic field causes a compass needle on Earth's surface to point toward the North Pole. But a bar magnet on your desk will change the direction of the compass needle. That must mean that the magnetic field of the bar magnet is stronger than the magnetic field of Earth at that point in space. How could a bar magnet produce stronger forces on the compass than Earth?

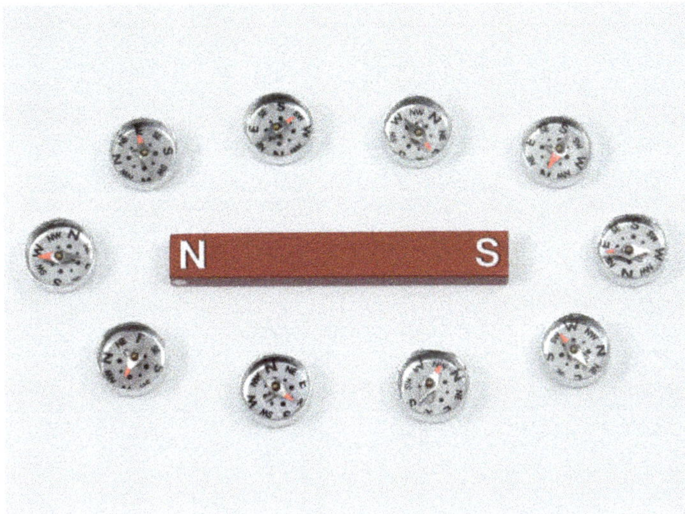

Earth's magnetic field is actually pretty weak on the surface where we live. That is because the magnetic field is created deep within the planet. Here on the surface of Earth, we are thousands of miles away from the outer core, the source of the field. Think back to what we figured out in Lesson 10. What happens to the magnetic force when we increase the distance from a magnet?

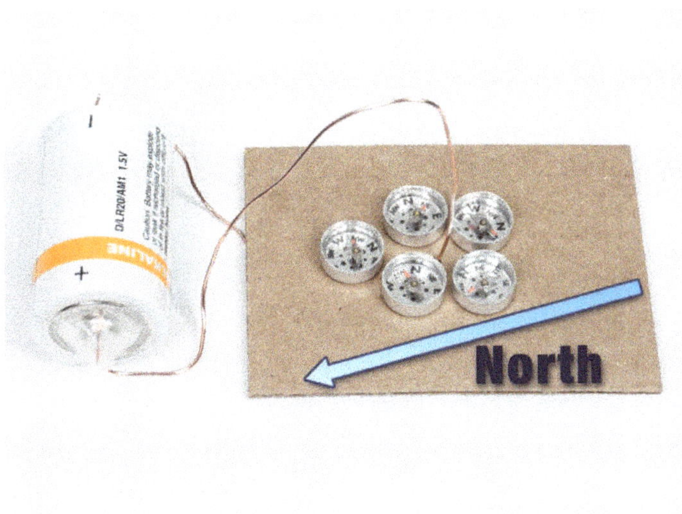

www.ingramcontent.com/pod-product-compliance
Ingram Content Group UK Ltd.
Pitfield, Milton Keynes, MK11 3LW, UK
UKHW052013041026
15590UKWH00008B/2317